T0255992

SpringerBriefs in Applied Sciences and Technology

SpringerBriefs present concise summaries of cutting-edge research and practical applications across a wide spectrum of fields. Featuring compact volumes of 50–125 pages, the series covers a range of content from professional to academic.

Typical publications can be:

- A timely report of state-of-the art methods
- An introduction to or a manual for the application of mathematical or computer techniques
- A bridge between new research results, as published in journal articles
- A snapshot of a hot or emerging topic
- An in-depth case study
- A presentation of core concepts that students must understand in order to make independent contributions

SpringerBriefs are characterized by fast, global electronic dissemination, standard publishing contracts, standardized manuscript preparation and formatting guidelines, and expedited production schedules.

On the one hand, **SpringerBriefs in Applied Sciences and Technology** are devoted to the publication of fundamentals and applications within the different classical engineering disciplines as well as in interdisciplinary fields that recently emerged between these areas. On the other hand, as the boundary separating fundamental research and applied technology is more and more dissolving, this series is particularly open to trans-disciplinary topics between fundamental science and engineering.

Indexed by EI-Compendex and Springerlink.

More information about this series at http://www.springer.com/series/8884

Madhusudan Singh

Node-to-Node Approaching in Wireless Mesh Connectivity

 Springer

Madhusudan Singh
Endicott College of International Studies
Woosong University
Daejeon
Korea (Republic of)

ISSN 2191-530X   ISSN 2191-5318 (electronic)
SpringerBriefs in Applied Sciences and Technology
ISBN 978-981-13-0673-0   ISBN 978-981-13-0674-7 (eBook)
https://doi.org/10.1007/978-981-13-0674-7

Library of Congress Control Number: 2018941980

Printed on acid-free paper

This Springer imprint is published by the registered company Springer Nature Singapore Pte Ltd.
part of Springer Nature
The registered company address is: 152 Beach Road, #21-01/04 Gateway East, Singapore 189721, Singapore

Preface

The IEEE 802.11s have newly become a scorching subject for researchers for the deployment of wireless networks. It is representing the connectivity of mesh networking in IEEE 802.11 amendment in static and ad hoc networks. However, it is Wireless Mesh Networks (WMNs) which often consists of mesh clients (laptop and cell phones), mesh routers, and gateways. WMNs have many attractive features such as highly reliable connectivity, easy deployment, self-healing, self-configuring, and flexible network expansion. Throughout the book, we have proposed two resolutions of routing and security with respect end-to-end delay, packet delivery ratio, and routing overhead in WMNs. For routing issue, we have proposed two routing mechanism, Novel Cluster-BasedNode-to-Node Approaching (NCBN-TO-NA) mechanism, which considers grouping between different clusters as Cluster Head (CH) and uses path discovery with the help of unicast method for very less congestion and packet loss, and Decentralized Hybrid Wireless Mesh Protocol (DHWMP). This protocol considers route announcement scheme of Mesh Portal Point (MPP) in WMNs.

For the evaluation of the proposed mechanism, we have used OPNET and NS-2 simulator and created a real-time test bed environment of WMNs in a Linux-based server. The book is intended for engineers, undergraduate, postgraduates, doctorate, and scientific degree applicants who carry out research, evaluation, and designing of hardware and software for secure communication in WMNs. The information contained in this book represents the results of the extensive work of the author with Prof. Sang-Gon Lee, Whye-Kit Tan, and Jun Huy Lam at the Dongseo University Busan, and Prof. Dhananjay Singh at Hankuk University of Foreign Studies, South Korea.

Daejeon, Korea (Republic of)   Madhusudan Singh

About the Book/Conference

This book highlights the routing protocols for Wireless Mesh Networks (WMNs; IEEE 802.11s). It provides an overview of the wireless networks (history, MANET, family of IEEE 802.11, WMNS, etc.) and routing protocols, such as AODV, DSR, OLSR, etc., and also highlights the two resolutions of routing protocols with respect to end-to-end delay, packet delivery ratio, and routing overhead in WMNs. Wireless Mesh Networks have become a hot topic for researcher into the deployment of wireless networks, and they represent the connectivity of mesh networking in IEEE 802.11 amendment in static and ad hoc networks. Moreover, WMNs have numerous attractive features, such as highly reliable connectivity, easy deployment, self-healing, self-configuring, and flexible network expansion. The book describes two routing mechanisms: Novel Cluster-Based Routing Protocols (NCBRP) and Decentralized Hybrid Wireless Mesh Protocol (DHWMP).

Contents

About the Author

Madhusudan Singh (SMIEEE) is currently an assistant professor at Endicott College of International Studies, Woosong University, Daejeon, South Korea. He has also served as research professor at Yonsei Institue of Convergence Technology (YICT) at Yonsei University, Korea (2016, June to 2018, Feb). Before joining YICT, he has worked as Senior Engineer at Research and Development department of Samsung Display, Yongin-Si, South Korea (2012.Mar. ~ 2016.Mar.).

He has received his Ph.D. in the Department of Ubiquitous IT, from Dongseo University (DSU) (Busan, South Korea) in Feb 2012. He has completed dual Master's, MS in the Department of IT with the specialization in Software Engineering from Indian Institute of Information Technology-Allahabad (IIIT-A), India in 2008, Jul., and MS in the Department of Computer Science from Uttar Pradesh (State) Technical University (UPTU), Lucknow, India in 2006, and BS degree in the Department of Computer Applications from Purvanchal University-Jaunpur, India.

Currently, he is Senior Member of IEEE (SMIEEE) and member of ACM, and many more research and scientific organizations. He is editor/reviewer/technical committee member of multiple IEEE, ACM and Springer international journals/conferences. He has published 50+ refereed research articles, 9+ national/international patents, and delivered 15+ technical talks as a speaker. His fields of research interests are Blockchain Technology, Cyber Security, Wireless Communication, Autonomous Vehicle, Artificial Intelligence, Internet of Things, Software Engineering, and Computer Vision.

Abbreviations

AODV	Ad hoc On-Demand Distance Vector Routing Protocol
BGP	Border Gateway Protocol
CBR	Constant Bit Rate
CSMA/CA	Carrier Sense Multiple Access/Collision Avoidance
CTS/RTS	Clear to Send/Request to Send
HWMP	Hybrid Wireless Mesh Protocol
IEEE 802.11s	WLANs with Mesh Network
MAC	Medium Access Control
MADWiFi	Multiband Atheros Driver for Wi-Fi (MADWI F I)
MANET	Mobile Ad hoc Network
MAPs	Mesh Access Points
MPPs	Mesh Portal Points
MWNs	Multi-hop Wireless Networks
NCBN-to-NA	Novel Cluster-Based Node-to-Node Approach Protocol
OLSR	Optimized Link State Routing
RERR	Route Error Message
RIP	Routing Information Protocol
RREP	Route Reply
RREQ	Route Request
STAs	Mobile Stations
TCP	Transmission Control Protocol
UDP	User Datagram Protocol
VAP	Virtual Access Point
WANs	Wireless Ad hoc Networks
WDS	Wireless Distribution System
WLANs	Wireless Local Area Networks
WMNs	Wireless Mesh Networks

List of Figures

List of Tables

Chapter 1
Wireless Networks: An Overview

1.1 Networks Mechanism

The network that uses wires is known as a wired (cable) networks. Initially the networks were mostly wired networks. The installation of networks has been a typical issue of the wired networks because the Ethernet cable should be connected to each and every computer. Definitely this kind of connection should be a time taking process. The wiring of a wired networks depends on lot of things like what kind of devices are being used in a wired network, whether the networks is using external modem or is it internal, the type of internet connection and many other issues.

A wireless network is a setup of communication networks among computers or other networks devices without using wires. Wireless networks are using radio waves for networks communication.

1.1.1 Wired Networks

The hardware implementation is a main task in the wired networks configuration. Once the hardware implementation is finished in a wired network, the remaining steps in a wired network do not differ so much from the steps in a wireless network. There are some things of wired networks that include cost, reliability and performance. In wired networks Ethernet cables, hubs and switches are very inexpensive. Some connection sharing software packages, like ICS, are free; some cost a nominal fee [1].

While making wired networks, Ethernet cable is the most reliable one because the makers of Ethernet cable continuously improving its technology and always produces a new Ethernet cable by removing the drawbacks of previous one. In terms of performance, wired networks can provide good results. Traditional Ethernet connection provides only 10 Mbps bandwidth, but now days 100 Mbps Fast Ethernet technology ready for use. Security in wired networks can be a little problem because wired Ethernet hubs and switches do not support firewalls. However, this problem

M. Singh, *Node-to-Node Approaching in Wireless Mesh Connectivity*, SpringerBriefs
in Applied Sciences and Technology, https://doi.org/10.1007/978-981-13-0674-7_1

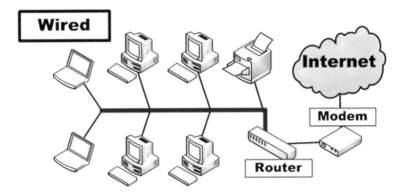

Fig. 1.1 Wired networks

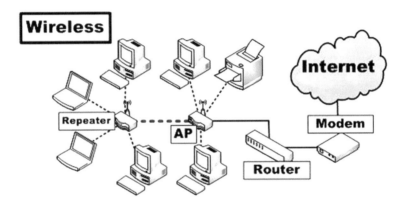

Fig. 1.2 Wireless networks

can be solved by installing firewall software on each individual computer in networks
[1]. In Fig. 1.1 is shown the wired networks.

1.1.2 Wireless Networks: An Overview

Wireless networks are sometimes known as Wi-Fi networks or WLAN. For wireless
networks, every node (communication devices) must need wireless adapter. In techni-
cally we are also known as IEEE 802.11 [2]. Figure 1.2 is shown the wireless network
architecture. The wireless networks are getting popular nowadays due to easy to setup
feature and no cabling involved. We can connect computers anywhere in communication range without the need for wires. Nowadays it uses the Wi-Fi [3], as a standard of
communication among different nodes. We are discussing details wireless networks
in others subtitles. In Fig. 1.3 we can see the Branches of wireless networks.

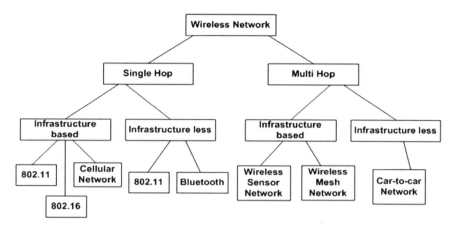

Fig. 1.3 Branches of wireless networking

We have discussed below in details about wireless networks types.

- *Point to point wireless networks*

The single hop communication is called point to point networks. In point to point networks, one node can communicate with single node at a time. It is just like a one to one function, where one node can make pair with another single node. This type networks is not good for big networks where speed and performance is main issue. In this kind of networks if one connection breaks the whole networks will stop working. It will cost more wire to build a networks and it is the most difficult networks in terms of configuring. This type networks are follows the tree and ring topology. In Fig. 1.4 is shown the example of point to point wireless networks [4].

- *Point to multi-point*

In this type networks, one node can communicate with more than one node. One node can control the communication process for all other node. If one node wants to communicate with other node, it must send data to the central node and central node will forward the data to that other node. This networks performance is based on quality of link between central device and all nodes. This type network is used star topology for communication [5]. Figure 1.5 is shown point to multi point networks.

- *Multipoint to multipoint wireless networks*

In these networks every node connected with all nodes. If one node wants to communicate with other node, it has many options to send data to the node and it will select feasible path to communicate with destination node. This type network is used mesh topology for communication. Nowadays this wireless network is gain very fame for communication technology. This type networks also called wireless LAN (IEEE802.11) networks [6]. Figure 1.6 is showing point to multi point networks.

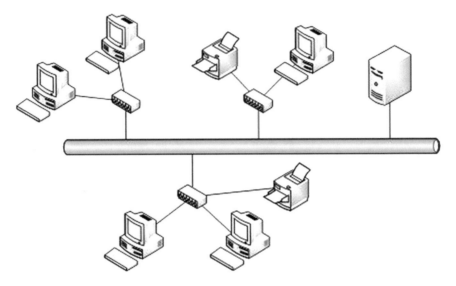

Fig. 1.4 Point to point wireless networks

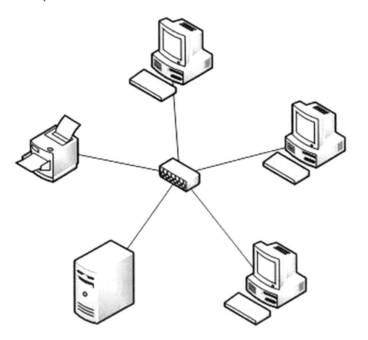

Fig. 1.5 Point to multipoint wireless networks

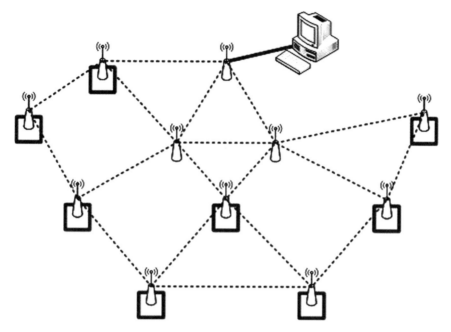

Fig. 1.6 Multipoint wireless networks

- *Classification of multi-hop wireless networks*

The classification of Multihop Wireless Networks (MWNs) has been done beauti-fully. To understand the concept of MWNs clearly, the figure is drawn [5]. MWN is a superset of Hybrid Wireless Networks (HWNs), Wireless Ad hoc Networks (WANs), Wireless Sensor Networks (WSNs) and WMNs. WANs have no infrastructure and posse's dynamic topology, WSNs are made of tiny sensor nodes and they can fol-low single hop wireless communication or Multihop wireless communication. On the other hand, HWNs can follow both single and multi-hop communications. In Fig. 1.7 we can see the classification of multi hop wireless networks.

- *Mobile ad hoc networks (MANET's)*

Mobile ad hoc networks (MANETs) are also type of wireless ad hoc networks that is why MANETs and WMNs are correlated with each other [1]. Even sometimes MANET is also called as mobile mesh networks and it can also be called as a wireless mesh networks. It is not necessary that mesh networks can also be mobile or wireless mesh networks. As previous diagram is showing that MANETs are infrastructure less but when the definition of WMNs is applied it is found that MANETs are the subset of WMNs as both of them are self-organizing. Also the study of different research papers reveals that WMNs and MANETs can be taken as same kind of net-works. So like WMN, MANET is also useful for larger coverage area like internet worldwide. MANET is made on the logic that each node is independent and free to

Fig. 1.7 Classifications of
MWNs

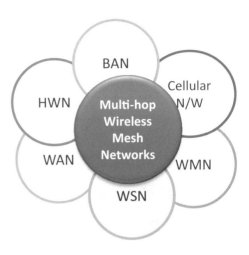

move in every direction. Since it works dynamically that is why while routing, routing protocols can easily find and update the selected paths dynamically. Moreover, MANET supports multi hop communication same like WMN [5]. The term ad hoc came from Latin language which means "for this purpose only". Since MANETs are very much in common with WMNs, the plus points of MANETs are also very much similar to WMNs. The communication is done through wireless links. Nodes that constitute these networks can perform the functions of routers and hosts. They use dynamic networks topology. Free of infrastructure like WMNs. It can be made at any place as it is a wireless network. The applications of MANETs are also same like WMNs. Nowadays United States (US) military has more interest in using MANETs. Information can be accessed easily as compared to wired networks. The main disadvantage of MANETs is that because of its wireless feature there are more chances of attacks on it. Attacker might attack easily to wireless networks as compared to wired networks. As in MANETs the nodes can communicate with their neighboring nodes also without the use of a central server, therefore when some node is affected or not working properly, it is hard to find that infected node as it has volatile networks topology. In Fig. 1.8 is shown the MANET networks.

When any networks have a large number of participants and needs a long observation period that time traces models provide accurate information. However, new networks environments, for example ad hoc networks, are not easily modeled if traces have not yet been created. In this type of situation, it is necessary to use synthetic models the second type, synthetic models, tries to show the behavior of mobile nodes (MNs) without the use of traces [6]. Realistic models for the motion patterns are needed in simulations in order to evaluate system and protocol performance. Mobility patterns have been used to drive traffic and mobility prediction models in the study of various problems of cellular systems, such as handoff, location management, paging, registration, calling time, and traffic load.

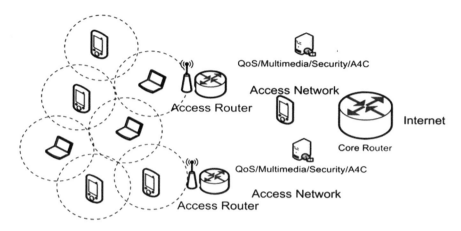

Fig. 1.8 Mobile ad-hoc networks

1.2 History of IEEE 802.11

802.11 are wireless local area networks (WLANs) which developed by the Institute of Electrical and Electronics Engineers (IEEE) working group. IEEE 802.11 is an emerging and very famous technology of computer communication [4]. IEEE 802.11 is a group of touchstones for implementing communication networks in different radio waves. IEEE 802.11 specifies an over—the air interface between wireless nodes. IEEE is accepted first 802.11 specifications in 1997. That specification has defined the media access control (MAC) and physical (PHY) layers for WLAN to wireless communication [6].

1.2.1 Family of IEEE 802.11

The installation of 802.11 families can do by two methods. First one is ad hoc mode. Ad hoc mode can provide communication basically peer to peer connection. The second method is infrastructure method. It allows as it allows wireless devices in a networks to communicate with an infrastructure. It is most required mode for the networks communication. Both modes are use wireless networks adapters termed as WLAN cards. IEEE 802.11 family has many standards amendment. Some amendment has described in below. Figure 1.9 shows IEEE 802.11 networks [6].

- **IEEE 802.11a**

The 802.11a amendment was accepted by IEEE in 1999. In this amendment researchers specifies radio transmit in three different radios transmission way 100 MHs unlicensed frequency bands in 5 GHz unlicensed National Information

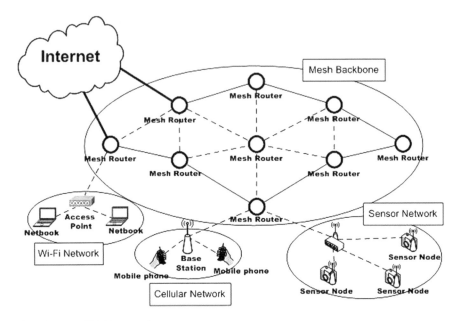

Fig. 1.9 IEEE 802.11 networks

Infrastructure range because this is less congestion band. This type device is supported up to 54 Mbps data rates. It uses OFDM method for modulation. This standard is newer than 802.11b. Its communication speed is better than 802.11b provides better speed [6].

- **IEEE 802.11b**

IEEE 802.11b was the oldest standard used in WLAN. 802.11b amendment was accepted by IEEE same year of 802.11a (1999). In this WLAN radios operate in the Unlicensed 2.4–2.4835 GHs Industrial Scientific and Medical (ISM) Band. The IEEE 802.11b devices are support up to 11 Mbps data rates.

- **IEEE 802.11g**

802.11g is the latest standard in the 802.11 family. The 802.11g amendment was submitted in 2003. The main aim was to achieve greater bandwidth but remain compatible to the 802.11 MAC. Here the Extended Rate Physical—OFDM (ERP-OFDM) is used as the modulation technique to achieve data rates of up to 54 Mbps. The 802.11g has removed the deficiencies of previous two standards. It provides better communication speed from other two standards.

Many others standards are present for WLAN techniques but other standards have different design goals like the 802.11s which has mobility and multi-hop design focus and capabilities. The family of standards all strives to solve specific issues in wireless networks.

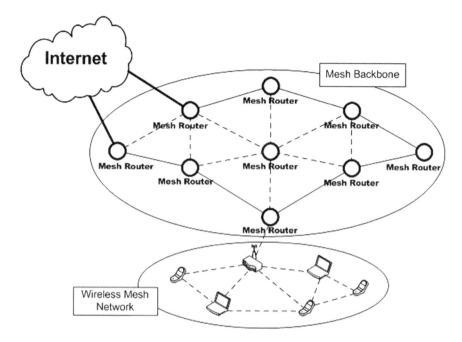

Fig. 1.10 IEEE802.11s wireless mesh networks

• **IEEE 802.11s** [7]

Wireless local area networks (WLANs) have become very popular in recent years. IEEE 802.11 is the standard for WLAN. This specification defines a physical layer and an Ethernet like MAC layer for wireless links [7]. IEEE 802.11 consists of mobile stations (STAs) and access points (APs). A mobile station is a networks device with a wireless networks interface card. APs are acting as bridge by providing connectivity to stations. APs are connected to each other through wired links.

Our research works is focused on IEEE 802.11s (Wireless Mesh Networks). We have discussed below about IEEE 802.11s. In Fig. 1.10 is shown an IEEE802.11s.

1.3 Differences Between Wireless Ad Hoc Networks and Wireless Mesh Networks

Wireless Ad hoc Networks (WANs) and Wireless Mesh Networks (WMNs) are very much similar with each other. Only problem is with the routing protocols. Those routing protocols that give best performance in WANs do not give reasonable performance when it comes to use in WMNs. Both of them use different protocols. They differ with each other in some matter which can be illustrated in following Table 1.1.

Table 1.1 Showing differences b/w WANs and WMNs

Issues	Wireless ad hoc	Wireless mesh
Networks topology	Highly dynamic	Relative static
Mobility of nodes	Medium to high	Low
Energy constraint	High	Low
Routing performance	Fully distributed on-demand routing	Fully distributed or partial distributed with table driven routing
Deployment	Easy to deploy	Planning required
Traffic characteristics	User traffic	User and sensor traffic
Relaying	By mobile nodes	By fixed nodes
Infrastructure requirement	Infrastructure less	Fully fixed or partial infrastructure

References

1. M. Singh, S.-G. Lee, D. Singh, H.-J. Lee, Impact and performance of mobility models in wireless ad-hoc networks. ICCIT 2009 International Conference on Computer Sciences and Convergence Information Technology, Korea, pp. 139–143 (2009)
2. IEEE P802.11s task Group IEEE Unapproved draft standard P802.11s/D4.0, Dec (2010)
3. M.S. Kuran, T. Tugcu, A survey on emerging broadband wireless access technologies. Comput. Netw. **51**(11), 3013–3046 (2007)
4. M. Singh, S.-G. Lee, H.-J. Lee, Performance mobility models for routing protocol in wireless ad-hoc networks. Korea Inst. Inf. Commun. Eng. **5**(9), 610–614 (2011) (Journal of Information and Communication Convergence Engineering). https://doi.org/10.6109/jice.2011.9.5.610
5. J. Bicket, D. Aguayo, S. Biswas, R. Morris, Architecture and evaluation of an unplanned 802.11b mesh network, the 11th annual international conference on Mobile computing and networking (MobiCom '05). ACM, pp. 31.42, USA (2005). http://doi.acm.org/10.1145/1080829.1080833
6. Z. Yan, L. Jijun, H. Honglin. *Wireless Mesh Networking, Architectures Protocols and Standard* (Auerbach Publications, USA, 2007)
7. I.F. Akyildiz, X. Wang, W. Wang, Wireless mesh networks: a survey. Elsevier Comput. Netw. J. **47**(4), 445–487 (2005)

Chapter 2
Wireless Mesh Networks Architecture

2.1 IEEE 802.11s Architecture

Wireless local area networks (WLANs) have become very popular in recent years. IEEE 802.11 is the standard for WLAN. This specification defines a physical layer and an Ethernet like MAC layer for wireless links [1]. IEEE 802.11 consists of mobile stations (STAs) and access points (APs). A mobile station is a networks device with a wireless networks interface card. APs are acting as bridge by providing connectivity to stations. APs are connected to each other through wired links.

IEEE 802.11 provides a cost effective and simple way for wireless networking. However, the problem is the wired connection between the APs. The wired links increases complexity and deployment cost in many situations. Therefore, it is desirable to connect the APs via wireless links as well and create a WLAN Mesh. In WMNs, APs turn into mesh access points (MAPs). Mobile stations are sometimes referred as mesh clients. The new IEEE 802.11s standard for WMNs introduces a third class of nodes called mesh points (MPs) [2]. MPs and MAPs support WLAN mesh services, allowing them to forward packets on behalf of other nodes to extend the wireless transmission range. Mesh clients can associate with MAPs but not with MPs. Mesh portals (MPPs) are MAPs which provides connectivity to other networks thus acting as a gateway for Mesh Networks.

2.2 Infrastructure of Wireless Mesh Networks

Infrastructure WMNs uses MPs and MAPs as backbone for the clients. Availability of gateway like MPPs provides connectivity to external networks such as Internet. The clients connect to the MAPs via standard 802.11, but do not forward packets [3]. This is the most used architecture nowadays. Access Points are also used for increasing coverage area. Figure 2.1 shows typical structure of this type of networks.

© The Author(s), under exclusive license to Springer Nature Singapore Pte Ltd., part of Springer Nature 2019
M. Singh, *Node-to-Node Approaching in Wireless Mesh Connectivity*, SpringerBriefs in Applied Sciences and Technology, https://doi.org/10.1007/978-981-13-0674-7_2

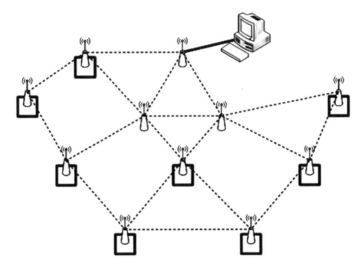

Fig. 2.1 Basic infrastructure WMNs

2.2.1 Client Wireless Mesh Networks

In this type of architecture, client nodes constitute the actual networks to perform routing and configuration functionalities as well as providing end user applications to customers. As an example let us consider a collection of MPs connected to each other, they can communicate to each other within networks and also forward data on behalf of other. In Client WMNs, a packet destined to a node in the networks, hops through multiple nodes to reach the destination [4]. Client WMNs are usually formed using one type of radios on devices. Moreover, the requirements on end-user devices is increased when compared to infrastructure meshing, since, in Client WMNs, the end-users must perform additional functions such as routing and self-configuration.

2.2.2 Hybrid Wireless Mesh Networks

This architecture is the combination of infrastructure and client meshing. Mesh clients can access the networks through MPs as well as directly meshing with other mesh clients. While the infrastructure provides connectivity to other networks such as the Internet, Wi-Fi, WiMax, cellular, and sensor networks; the routing capabilities of clients provide improved connectivity and coverage inside the WMN. The hybrid architecture will be the most applicable case in our opinion. IEEE 802.11s is the typical scenario of Hybrid WMNs, we have clients (Ordinary stations), which uses Mesh Access Points (MAPs) for accessing networks and we have independent nodes (MPs) which can directly access the networks [5]. All these entities are based on

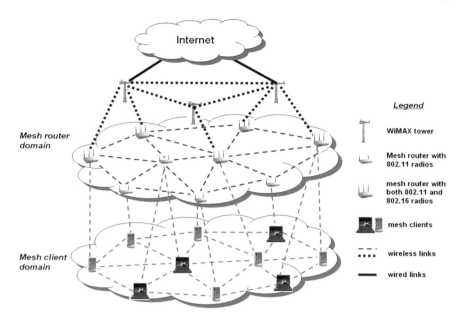

Fig. 2.2 Hybrid WMNs

802.11-based radio technology. The typical structure of hybrid WMNs are shown in Fig. 2.2.

Due to different system architecture mesh networks have different requirement to the physical layer, the MAC mechanism and the routing protocol than legacy IEEE 802.11 LANs [1]. The changes made in above layer for WMN are discussed in next section. In 802.11s which employ multi-hop capabilities, data can be routed along an alternate path to avoid interference. Also if a node requires a large amount of resources like in the case of single hop access points where bottlenecks are created, the networks can dynamically route traffic through other networks nodes hence avoiding the congested node.

In the WMN where routing is done in layer 3 cooperation is only done between stations that are mesh enabled. Hence stations without mesh capabilities do not join in the networks. The 802.11s provides an extension to the MAC frame header where APs could be able to connect to each other wirelessly establishing peer to peer wireless links creating a wireless distribution system (WDS) or a Mesh Basic Service Set (MBSS) [5]. This would create a backhaul infrastructure with no need of all the APs being connected to an 802.3 networks and therefore merging the lines between infrastructure and client devices in some usage scenarios. The extensions in the MAC header of 802.11s frame makes it possible for addresses of legacy stations to be reached in a WMN. Also provided for in the standard is a robust and transparent data layer which supports all higher protocol layers. This was necessary so that the new devices could be compatible with the existing ones hence the only change being in the MAC layer header frame. IEEE 802.11s aligns with the 802.11i amendment which concerns security and also 802.11e which concerns quality of service (QOS) [6].

References

1. IEEE, P802.11s draft d3.02, draft amendment to standard IEEE 802.11: ESS mesh networking, standard, (2008)
2. D.B. Johnson, D.A. Maltz, J. Broch, *"DSR: The Dynamic Source Routing Protocol for Multi-Hop Wireless Ad Hoc Networks" Ad Hoc Networking, chap. 5*, (Addison-Wesley, USA, 2001), pp. 139–172
3. A. Sgora, D.D. Vergados, P. Chatzimisios, *"IEEE802.122s Wireless Mesh Networks: challeneges and Perspectives", Mobile Lightweight Wireless* (Springer, Berlin, 2009), pp 263–271
4. X. Wang, A.O. Lim, IEEE 8002.11s wireless mesh networks: framework and challenges. J. Elsevier (2007)
5. A. Al-Saadi, R. Setchi, Y. Hicks, S.M. Allen, Routing protocol for heterogeneous wireless mesh networks. IEEE Transac. Veh. Technol. **65**(12), 9773–9786 (2016). https://doi.org/10.1109/tvt.2016.2518931
6. C.E. Perkins, E.B. Royer, Ad hoc on-demand distance vector routing, in *IEEE Workshop on Mobile Computing Systems and Applications* (1999), pp. 90–100

Chapter 3
Routing Protocol for WMNs

3.1 Introduction

WMNs require a different classification of routing protocols. In WMNs routing protocols are classified as either proactive or reactive. As the name suggests proactive routing protocol keep the routes before one is needed. It tries to keep up to date information. Reactive protocol finds a route only when a node wants to communicate with another. Hybrid routing protocols are combination of the above two. Some routes to the destination are maintained proactively and other are created on demand [1].

Two major attributes of routing protocols are efficiency and convergence time. Efficiency is the share of routing traffic in overall traffic. Proactive routing protocols are efficient in high load scenarios while reactive protocols work well in low traffic settings. Convergence time refers to the time span needed to have correct routing tables after a topology change. Link state protocols usually have a better convergence time than distance vector protocols [2].

Routing Protocols in WMNs has few differences as compared to conventional protocols. Hybrid Wireless Mesh Protocol (HWMP) which is the default protocol for IEEE 802.11s operates on MAC layer. This is because MPs and MAPs don't support IP addresses [3]. Other reason may be that multi-radio routing for capacity improvement and multi-path routing for load balancing and fault tolerance can be desirable in WMNs.

3.2 Ad Hoc On-Demand Distance Vector Routing (AODV)

AODV is hop-by-hop routing protocols developed for wireless Ad hoc networks. It offers quick adaptation to dynamic link conditions, low processing and memory overhead. When a host wants to find a route to a destination it broadcast a route request (RREQ) message. The RREQ contains addresses (source and destination), sequence number and a broadcast identifier [4].

Nodes other than destination receiving RREQ message either re-broadcast or respond with route reply (RREP), depending on flags setting in RREQ message.

© The Author(s), under exclusive license to Springer Nature Singapore Pte Ltd., part of Springer Nature 2019
M. Singh, *Node-to-Node Approaching in Wireless Mesh Connectivity*, SpringerBriefs in Applied Sciences and Technology, https://doi.org/10.1007/978-981-13-0674-7_3

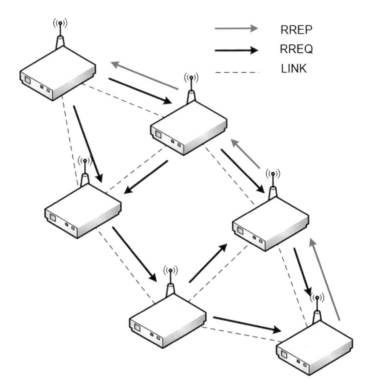

Fig. 3.1 AODV route discovery

When forwarding a RREQ node stores broadcast identifier, source address and maintains a reverse route. In order to avoid loop, RREQ are re broadcasted only when a request with the same source address and broadcast identifier has not been processed before. Concept of sequence number is used for route updating. Thus an intermediate host replies with a RREP when it has a fresh enough route to the destination. RREQ message was broadcasted by source node. Intermediate node creates and maintains a reverse route to the source node. Destination node, on receiving RREQ sends a unicast RREP to the source node on the same path that was created during RREQ [3]. Figure 3.1 is an example of AODV routing.

3.3 Dynamic Source Routing (DSR)

DSR is an efficient routing protocol designed for Ad Hoc networks. Intermediate nodes are used to forward packets over multiple hops between nodes not directly within transmission range of each other. The protocol is composed of two mechanisms of Route Discovery and Route Maintenance, which work together to allow nodes to discover and maintain source routes to arbitrary destinations in the Ad Hoc networks. In DSR if the source node which wants to communicate with a certain destination has the route to destination in its cache, it will insert the route in data packet headers. If source nodes don't have the route information it will start a route discovery process. In this process RREQ packets are flooded in the networks until these packets reach the destination. Destination node on receiving the first route request will send the RREP message to the source node. In case on link breakage RERR message is sent the destination and destination tests another route. Therefore, only after making error in current route, destination seeks another route. This mechanism causes delay in packet delivery [5]. The protocol allows multiple routes to any destination and allows each sender to select and control the routes used in routing its packets.

3.4 Optimized Link State Routing (OLSR)

OLSR is a proactive protocol based on table driven techniques. The proactive nature of the protocol makes it to maintain routes to all nodes in the networks by frequent exchange of topology information with other nodes in the networks. OLSR adopts the Multipoint Relay (MPR) mechanism to reduce the routing overhead caused by the flooding of control information in the networks. Each node in the networks selects a set of one-hop symmetric neighbor that can cover at least two-hop symmetric neighbor as MPR. It is only the selected MPR that is permitted to forward a control message across the networks, thereby limiting the number of nodes in the networks to retransmit control message. This approach significantly reduces the routing overhead in the networks. The control message contains the list of link state information of the originating nodes, which the MPR declares periodically to all its MPR selectors. OLSR floods only the partial link state to enable shortest path routes and operates in a distributed approach without need for central entity. The local link information on each node and the link state information received from the MPR are used in route calculation from a given source node to any reachable destination in the networks. Its reaction to change in topology can be optimized by reducing the time interval for the periodic control [4]. OLSR is suitable for large and dense networks where large subset of nodes exchange traffics between each other. Each control message in OLSR contains sequence number to distinguish fresh information from stale information and does not require sequenced delivery of information. OLSR also support protocol extension such as sleep mode operation, multicast routing etc. There are fundamental message types that must be observed in the OLSR implementation and must maintain compatibility with old implementation if any additional message type is to be implemented in OLSR [5]. The messages are listed below.

- Hello Message: It is used for conducting link sensing, neighbor detection and MPR signaling process.
- Topology Control Message: this is used for the topology declaration (advertisement of link state).
- MID Message: It handles the declaration of the presence of multiple interfaces on a node.

3.5   Extensible Mesh Routing Framework

The IEEE802.11s working group (TG) has defined an extensible mesh routing framework to ensure a common standard that will harmonize all WLAN mesh networks devices from various vendors. The framework brought the routing in the upcoming IEEE802.11s on Layer 2 of the OSI reference model using the Media Access Control (MAC) address and radio—aware metrics [4]. This will be the first WMN technology to route on layer 2. IEEE802.11s is capable of forming a transparent 802 broadcast domain that support any upper layer protocols. This approach is in contrast with the routing on Layer 3 of the OSI model with Internet Protocol (IP) address designed for MANET. Routing on layer 2 is termed path selection in WLAN mesh networks to distinguish it from Layer 3 routing. The mesh path selection describes the selection of single hop or multi hop path and the mesh forwarding describes the forwarding of multi hop action frame and data across these paths between Mesh Stations (STAs) at the data link. The standard supports unicast, broadcast and multicast communication. The extension in the mesh routing framework permits flexible implementation of different path selection protocols and path selection metrics. This gives IEEE802.11s the opportunity to be used in different environment according to vendor specific application need. The extensible framework has below mentioned characteristics for node to node approaching in wirless mesh connectivity:

- Default mandatory path selection protocol, Hybrid Wireless Mesh Protocol (HWMP) and default mandatory path selection metrics (Airtime link metric) which every IEEE802.11s devices must capable of using to enable interoperability of devices from different vendors [6].
- An optional path selection protocol, Radio—Aware Optimized Link State Routing (RA-OLSR).
- Vendor specific path selection protocol and or path selection metrics for special application need.

The standard recommends that a path selection protocol and a path selection metric shall be implemented at a time on mesh STA.

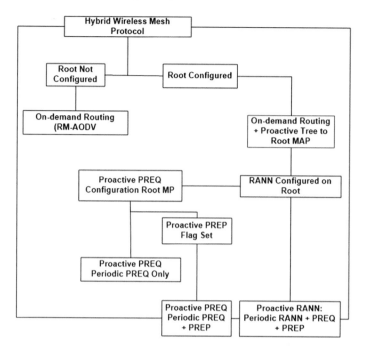

Fig. 3.2 Hybrid wireless mesh protocol (HWMP) [6]

3.6   Hybrid Wireless Mesh Protocol (HWMP)

The information provided in this section is based on the Pre IEEE802.11s draft standard D2.07 [4], 2009. Hybrid Wireless Mesh Protocol is the default path selection protocol developed for use in the upcoming IEEE802.11s standard. Every WLAN mesh networks devices from various vendors shall use HWMP and airtime link metric to ensure interoperability. The protocol combines the strengths of both the proactive and reactive routing protocols due to its hybrid nature. HWMP development is based on AODV specifications [6]; it was redesigned to use MAC address and radio aware path selection metric for path selection at the data link (Fig. 3.2).

References

1. R. Bruno, M. Conti, M. Nurchis, Opportunistic packet scheduling and routing in wireless mesh networks, Wireless Days (WD), 2010 IFIP, pp. 1–6, 20–22 (2010). https://doi.org/10.1109/wd.2010.5657736
2. A.H. Omari, A, H, Khrist, A dynamic and Reliable Mesh routing Protocol for Wireless Mesh Networks (DRMRP). IJCNS Int. J. Comput. Sci. Netw. Secur. **9**(4) (2009)
3. T. Clausen, P. Jacquet, Optimized Link State Routing Protocol (OLSR), IETF RFC 3626 (2003)

4. R. Bruno, M. Nurchis, Survey on diversity-based routing in wireless mesh networks: challenges and solutions. Comput. Commun. **33**(3), 269–282 (2010)
5. Y. Zuo, Z. Ling, Y. Yuan, A hybrid multi-path routing algorithm for industrial wireless mesh networks. EURASIP J. Wirel. Commun. Netw. **2013**(82) (2013). http://jwcn.eurasipjournals.com/content/2013/1/82
6. M. Bahr, Update on the hybrid wireless mesh protocol of IEEE 802.11s. IEEE International Conference on Mobile Ad-hoc and Sensor Systems, Pisa, Italy (2007)

Chapter 4
Novel Cluster Based Node to Node Approaching (NCBN-to-NA in WMNs Process) in Wireless Mesh Connectivity

4.1 Introduction of Novel Cluster Based Node to Node Approaching (NCBN-to-NA) in Wireless Mesh Connectivity [2]

We design a cluster based node to node approaching (routing) scheme for WMNs as we have mentioned, most of the algorithm use initially broadcast route request for the entire node in networks [1], [2]. It is possible that if networks are partitioned into the cluster, we can reduce the initial broadcast to all nodes. As each cluster has one head that have all the information of its neighbor, so path request is multicast to different cluster heads only [3]. In this scheme we distribute the whole mesh networks into groups (Fig. 4.1). Mesh point portal (MPP) assigned one node as a cluster head (CH) of each cluster group and stored the cluster head information in its own table such as CH id, CH neighbors etc. Figure 4.1 shows cluster group in WMNs.

4.2 NCBN-to-NA in WMNs Process [2]

4.2.1 Path Establishments

- **Setup of cluster head**

Each cluster head has some extra authority compare to others cluster member. Each cluster head has two tables, one table stores the information of neighbors cluster heads and second table stores the information about cluster group members which is assigned by the MPP. Every cluster member stores the information of his CH.

- **Path creation**

When a normal cluster member wants to communicate with any destination node, it sent the path request (PREQ) message to his cluster head (Fig. 4.2a).

© The Author(s), under exclusive license to Springer Nature Singapore Pte Ltd., part of Springer Nature 2019
M. Singh, *Node-to-Node Approaching in Wireless Mesh Connectivity*, SpringerBriefs in Applied Sciences and Technology, https://doi.org/10.1007/978-981-13-0674-7_4

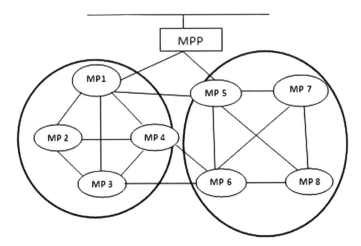

Fig. 4.1 Wireless mesh networks divided into cluster groups

Then cluster head checks its own group member list. If the destination exists in the same group, it sends path reply with path information quickly and source node starts transmission according to that path. If destination node belongs to other cluster, cluster head sends PREQ message to mesh portal (Fig. 4.2b) and the mesh portal multicast PREQ message to all cluster heads (Fig. 4.2c).

Cluster heads check own group table and if any CH finds destination node in area then its send PREQ message to destination node (Fig. 4.2d). Destination node sends own status to his cluster head (Fig. 4.2e).

- **Path reply**

After received status message of destination node, destination cluster head sends path reply message to mesh portal (Fig. 4.3f), and send them to source node with destination path information. MPP forwards PREP, which is received from destination CH2 to source CH1 as shown in Fig. 4.3g. CH1 forwards.

- **Path setup**

A bidirectional path established between the CH1 & CH2 nodes shown in (Fig. 4.3h) and it forwards the destination information to source node (Fig. 4.3i). So the final path between the source and destination nodes is as showing (Fig. 4.3j).

In this scheme, MPP multicast during path discovery for only once and remaining all transmission use unicast messaging. Hence it reduces power consumption and improves the networks performance Mesh portal and cluster head periodically updates own table that helps to detect any change occurs in networks. Further, in case of link-failure, transmission redirected to mesh portal by CH of the clusters that eliminate error message (RERR) broadcasting, but when cluster head fail, then broadcasting is required.

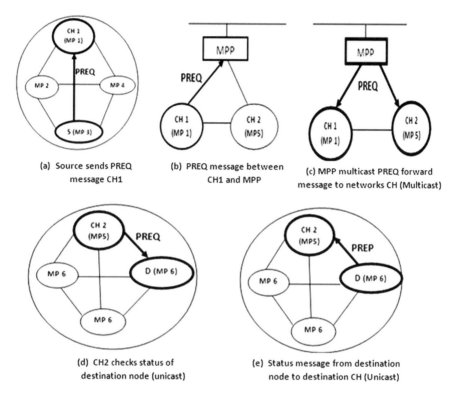

(a) Source sends PREQ
message CH1

(b) PREQ message between
CH1 and MPP

(c) MPP multicast PREQ forward
message to networks CH (Multicast)

(d) CH2 checks status of
destination node (unicast)

(e) Status message from destination
node to destination CH (Unicast)

Fig. 4.2 Path discovery process in NCBN-to-NA

4.3 Performance Analysis of NCBN-to-NA

In our simulations, we have modified the original 802.11s to design an improved
contention based NCBN-to-NA protocol to perform channel switching in a multi-
channel single radio environment. We improve on this environment at the sink node,
where previously we observed that channel switching among nodes by a single radio
at the sink node causes severe degradation [3].

In addition, the mesh routers with IP gateway functionalities were assigned as
backbone routers. In the model, 18 traffics sources are assumed to have maximum
mobile average speed of 5 m/s. 25 mesh IP gateway routers are placed in multiple
administrative domain (AD) of 100 nodes clients mesh. The analysis is done for
the traffic packets transmissions. The novel cluster based routing protocol has been
optimized for the traffics in the modeled simulation environment and the performance
of proposed routing protocol has been evaluated with respect to existing routing
protocols. The hierarchal architecture of the WMN was implemented using AD
cluster design in a hierarchal structure. In the model, every IP router was dynamically
assigned as AP for each of the four WLAN mesh nodes in the cluster-base pattern.

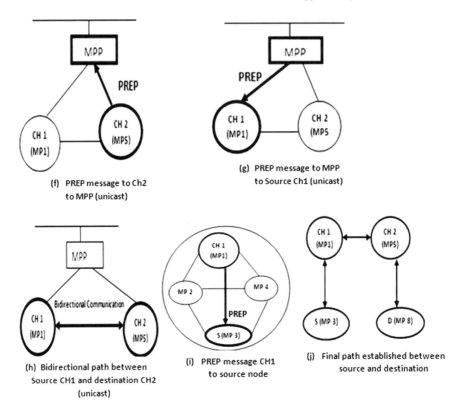

Fig. 4.3 Path reply (PREP) process in NCBN-to-NA [2]

The IP configuration and dynamic assignment of node addresses are stores in the network. While the Cluster-based attributes are enabled in the advanced attributes of the OPNET simulation modeler together with the IP routers gateway functionalities. These Cluster-based attributes are used to activate traffic engineering techniques and flows of transmissions [4].

These wireless traffic scenario adaptations were enabled in the mesh client nodes of the WMN model. In addition, the IP gateway functionalities were enabled in the routing protocol advanced attributes, mesh routers, AP for backbone route transmission and the inter-domain communication. In Table 4.1 show parameters for simulation in Opnet simulator.

In the WMN environment, we have analyzed the performance at the sink of the NCBN-to-NA protocol by simulations with OPNET [2–5]. A different simulation scenario has been studied according to four different performance metrics: packet delivery ratio, end-to-end delay, aggregate throughput, hope count. The WMN (nodes) are randomly placed and to analyze the channel capacity of WMNs there were 10 number of cluster heads placed in a 1000×1000 m^2 area. The radio range is set to 50 m. The radio bandwidth is 2 Mbps. The number of nodes is 100 (Fig. 4.4).

Table 4.1 Parameters for NCBN-to-NA simulations

Parameter	Value
Simulation area	$1000 * 1000 \, M^2$
Total no. of nodes	100
Total no. of CH	10
Simulation time	200 s
Packet size	512 bytes
Bandwidth	2 Mbps
Traffic type	CBR
Packet rate	10 pkts/s

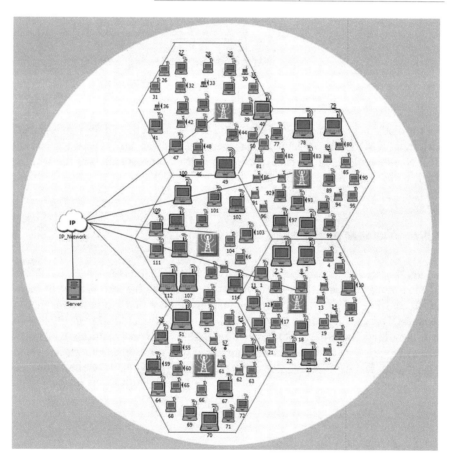

Fig. 4.4 OPNET modeler for cluster-based routing environment [2]

To evaluate the performance of proposed NCBN-to-NA mechanism, we have to analyze the capacity of each cluster head (CH) and the capacity of the whole network.

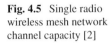

Fig. 4.5 Single radio wireless mesh network channel capacity [2]

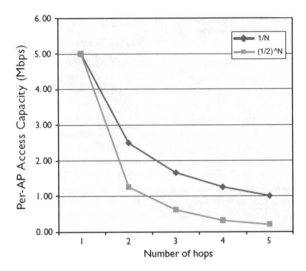

For the analysis performance, we have used single radio channel for wireless mesh networks (WMNs). As we know, the average channel capacity is between 1/N and (1/2) N. Where, N is the number of wireless hops in longest path between wireless single node and wired networks (Fig. 4.5).

4.3.1 Number of Path Discovered

In Fig. 4.6 represents average path discovered in NCBN-TO-NA and AODV and OLSR routing protocol. While AODV can discover the highest number of paths, OLSR and NCBN-TO-NA discover a smaller number due the proactive mechanism. AODV uses broadcast message and OLSR has predefined path method for path discovery. But NCBN-TO-NA routing uses unicast and multicast method for path discovery. Due to this it has less number of path discovery features. Each routing protocol has its own pros and cons, depending on the network conditions and requirements.

4.3.2 Packet Delivery Ratio

Packet delivery ratio (PDR) denotes the ratio of the total packets received by the destination to the total packets sent by the source during the whole simulation, which reflects the efficiency of the routing. The optimal number of paths in a multipath solution by measuring delivery ratio in different circumstances. The delivery ratio is also used (in addition to the number of paths found in each case) to determine the

Fig. 4.6 Average numbers of paths discovered by each protocol

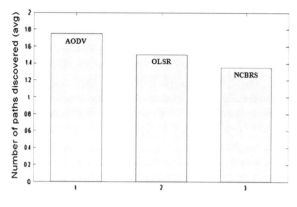

Fig. 4.7 Packet delivery ratio as a function of number of nodes

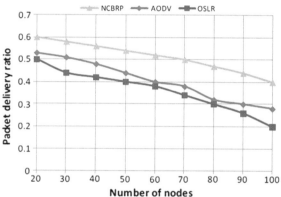

suitability of different clustering algorithms to support multi-path routing. The more the packets received, the better the performance.

That is

$$PDR = \frac{\Sigma \Pr_d}{\Sigma Ps_s}$$

where PDR denotes the packets delivery ratio during the whole simulation and $\Pr_d$ denotes the packets received by the destination and Ps_s denotes the packets sent by the source.

Figure 4.7 shows the packet delivery ratios for AODV, OLSR and NCBN-TO-NA with respect to the number of nodes. We can see that NCBN-to-NA performs better than OLSR and AODV in high or low network size. PDR of all the three routing protocols decreases when increases number of nodes. AODV performs better than OLSR in number of nodes. NCBN-TO-NA delivers almost (0.4–0.6%) packet in 20–100 nodes. We can see that NCBN-TO-NA's (0.4%) performance is much better than OLSR (0.2%) and AODV (0.28%).

4.4 Discussion

In this section we have described NCBN-to-NA protocol. In this protocol, the networks have partitioned into number of clusters. If the destination is within the cluster, cluster head sends path information to the source, as unicasting message. If the destination node is not a member of same cluster as source, then the cluster head send route request to the MPP as source PREQ, and it send route request to all cluster head through multicast. So again it does not require broadcasting of route request for path. Larger the cluster size less will be the number of multicast and more unicast, but it's an overload on cluster head. Decision of size of cluster and number of cluster depend on the topology and application of networks.

We have shown the performance of the NCBN-to-NA with respect to AODV and OLSR. We have shown that, NCBN-to-NA has given better performance for channel capacity, path discovery, packet delivery ratio, end to end delay except routing overhead. NCBN-to-NA is good routing protocol for WMNs compare to AODV and OLSR.

4.5 Summary

The section has presented routing scheme for WMN, aimed to reduce the usage of broadcast message for path discovery and the failure of path message (Path Error) over the networks to increase the efficiency of networks. Nodes were divided into clusters, and each cluster has cluster head. Hence, request for path is sent to cluster heads using multicast message only, through MPP. This eliminates the necessity of broadcast that reduces power consumption and improves networks performance. Further, any link failure would be handling by MPP as well and no broadcasting of error message is required except cluster head failure.

References

1. S. Ghannay, S.M. Gammar, F. Kamoun, Comparison of path selection protocols for IEEE802.11s WLAN Mesh Networks. IFIP International Federation for Information Processing, wireless and mobile Networking (Springer, Boston) (2008), pp 17–28
2. M. Singh, S.G. Lee, D. Singh, A simulation-based performance analysis of a cluster-based routing scheme for wireless mesh networks. Int. J. Multimedia Ubiquit. Eng. **8**(5), 283–296 (2013). ISSN No: 1975-0080
3. M.E.M. Campista, P.M. Esposito, I.M. Moraes, L.H.M.K. Costa, O.C.M.B. Duarte, D.G. Passos, C.V.N. Albuquerque, D.C. Muchaluat-Saade, M.G. Rubinstein, Routing metrics and protocols for wireless mesh networks. IEEE Netw. **22**(1), 6–12 (2008)

4. M. Singh, S.G. Lee, Group mechanism for wireless mesh network routing protocol to future internet services. The 2012 FTRA International Conference on Advanced IT, engineering and Management (AIM 2012), Seoul, South Korea, Feb 2012
5. M. Singh, S.G. Lee, T.W. Kit, L.J. Huy, Cluster based routing scheme for wireless mesh networks. The 13th International Conference on Advanced Communication (ICACT-2011), IEEE Communications Society, Phoenix Park, South Korea, 13–16 Feb 2011

Chapter 5
Decentralized Hybrid Wireless Mesh Protocol (DHWMP) Mechanism

5.1 Introduction

In decentralized hybrid wireless mesh node to node connectivity approach, we have modified the responsibility of mesh portal point (MPP), make some modifications in root announcement, modify proactive routing protocol, and try to solve the problems of the reactive routing protocol. That has used less energy and path setup time because in our scheme, we used 1 multicast and 3 unicast message for path discovery process. Due to this reason it will create fast path between source node and destination. Wireless mesh networks have hybrid wireless mesh protocol (HWMP) default routing protocol [1]. HWMP has used both reactive and proactive routing protocol. In path discovery process it broadcast the PREQ message in networks. Message broadcast is needed more power and energy and sometimes congestion also occurs in the networks.

Therefore, the existing HWMP is a very good protocol for the WMNs, but it has some drawbacks such as centralized root and tree-based communication in proactive routing [2]. If the shortest path exists between the source and the destination even, then the node still has to follow the root. Moreover, there is a drawback with the reactive routing: it is initiated with the PREQ. We have modified the exiting PREQ information element to provided DHWMP to support HWMP issues.

5.2 Default Routing Protocol HWMP

The IEEE 802.11 task group "S" (TGs) first meeting on WLAN mesh networking was held in July 2004 [3]. The goal of the group is to develop an extensible standard amendment for WMNs based on IEEE802.11 [3], which can be used in a flexible manner for many usage scenarios.

The IEEE802.11s draft can be split into four parts—routing, MAC enhancements, security, and general IEEE802.11-related topics. The IEEE802.11s draft standard D1.06 describes a default routing protocol HWMP [4]. Every IEEE802.11s compliant device must implement HWMP in order to use it.

© The Author(s), under exclusive license to Springer Nature Singapore Pte Ltd., part of Springer Nature 2019
M. Singh, *Node-to-Node Approaching in Wireless Mesh Connectivity*, SpringerBriefs in Applied Sciences and Technology, https://doi.org/10.1007/978-981-13-0674-7_5

A mesh point (MP), which is usually a mesh portal, can be configured to broadcast mesh portal announcements, which sets up a tree with the mesh portal as root of the tree [4]. One of these mesh portals that periodically broadcast mesh portal announcements becomes the designated root mesh portal by configuration or a selection process. This tree allows proactive routing through mesh portals. The foundation of HWMP is an adaptation of Ad-hoc On Demand Vector [AODV] to radio aware link metrics and MAC addresses. The on-demand path setup is achieved by a path discovery mechanism that is very similar to the one of AODV. If a MP needs a path to a destination, it broadcasts a path request message (PREQ) into the mesh networks. MPs will rebroadcast the updated PREQ whenever the received PREQ corresponds to a newer or better path to the source. Similarly, the requested destination MP will respond with a path reply message [5]. Reactive routing always initiates PREQ step, even when the path is already found and stored in the routing table of source node. Intermediate MPs that have already a valid path to the requested destination can respond with a path reply (PREP). Sometimes, the routing technique faces some errors during data transmission. If a link breaks when the route is active, the upstream of the break propagates. Then, a path error (PERR) message will be send to the source node to inform the unreachable destination [4]. Subsequently, the source node stops data transmission.

The proactive component of HWMP is the extension with a proactive routing tree to specially designated MP. Because proactive routing has a centralized and fixed common tree root, it always uses the root path even though there are other short paths, which do not include the root node. This leads to bottleneck problem and resource wastage.

5.3 Decentralized Hybrid Wireless Mesh Protocol (DHWMP) [4]

In proactive routing, source MP generates PREQ for destination MP and sends it to MPP. Source node maintains a neighbor id and destination id in its own routing table. Neighbor id is the id of a node that one hop distance away from the source node; destination id is the id of a node to which the source node wants to communicate.

The source node will reply to the PREQ received from MPP. MPP will keep updating its table after the addition or deletion of any MP. On the deletion or addition of any MP from the networks, its own table is immediately updated.

MPP is fixed point in the networks, and maintains two routing tables. The first routing table stores information on PREQ, such as source MP address, source id, destination id, hop count, and maintains the Neighbor query. After receiving PREQ from source node, MPP will generate a Neighbor query to the entire networks in order to obtain the neighbor information. In the second routing table, the MPP stores the source path details such as source MP, destination MP, intermediate MPs, and root ids. Root is found based on the calculation of neighbors' information. The common

MP between source and destination is assigned as a root, and MPs that compose the shortest path are assigned as intermediate MPs. The root will change according to the source MP and destination MP. The second table maintains the complete information about the root and path for the future use. If any node is added to the network or deleted from the network, then the MPP will update its table. The Modified proactive PREQ mechanism is defined below.

5.3.1 Path Request (PREQ) Mechanism

When a source MP S wants to send data to a destination MP D, e S checks its routing table to find whether any path exists to D. If not, then S discovers a new route for D. S is also known as the originator.

In this scheme, PREQ find the route through two phases. In the first phase, the source MP sends the PREQ to the MPP. In the second phase, MPP broadcasts the Neighbors query in the networks and finds the common and nearest neighboring MPs between the source and the destination. Figure 5.1 shows the information element (IE) present in PREQ.

- Element ID—To be define (TBD)
- Destination address—Address of the destination MP for which the source node has issued the PREQ
- Source Address—Address of originator MP
- MPP Address—MPP address contains the address of that MPP, which was sent by the source PREQ to the destination MP
- Hop count—The number of hops from the originator to the MP transmitting the Request

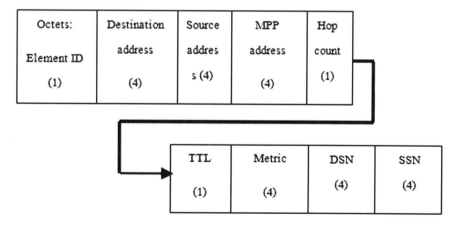

Fig. 5.1 PREQ element fields send by source node to MPP

Element ID (1)	Length (1)	MPs ID (2)	Hop count (4)	TTL (1)

Fig. 5.2 PREQ element fields send by neighbors node

- TTL—Time to live is the maximum number of hops allowed for this Information Element [IE]
- Metric—The cumulative metric from the originator to the MP transmitting the PREQ
- DSN—Destination sequence number (DSN) field has the sequence number of the destination
- SSN—Source sequence number (SSN) field has the sequence number of source MP.

After receiving the proactive PREQ from S, the MPP stores all information about the source, such as destination id, source id, and neighbors' id, and then it generates a Neighbor PREQ "neighbor's id of each node in networks". This query is then broadcasted to all the nodes. The IE of neighbors PREQ is shown in Fig. 5.2.

- Element ID—TBD
- MP id—This field we store the id of that MP that sends the neighbors information to MPP
- Hop count—The number of hops from the originator to the MP transmitting the Request
- TTL—Time to live is the maximum numbers of hops allowed for this IE.

5.3.2 Path Reply (PREP)

On receiving the neighbor query from MPP, each MP sends the information of its neighbors to the MPP, which is maintained in the routing table of MPP. It searches for a common feasible neighbor between the source and destination and define that particular node as a root. The complete information about the source and destination is then sent to that particular node so that it can prepare itself to function as a root. MPP sends the neighbor query and receives the neighbor information from all the existing MPs. Figure 5.3 shows the information elements of neighbors query PREP.

- Element ID—TBD
- Length—Length of the IE
- Neighbors Address—Neighbors' address information.

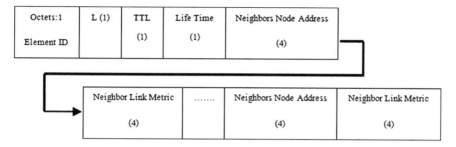

Fig. 5.3 PREP with neighbor's information for MPP

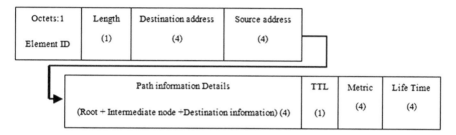

Fig. 5.4 PREP by MPP to source node with path information [3]

- Neighbors link Metric—The cumulative metric from the originator to the MP transmitting the PREQ.
- TTL—Time to live the maximum numbers of hops allowed for this IE
- Life Time—The time for which MPs receiving the PREQ consider the forwarding information to be valid.

   Figure 5.4 shows the IE of MPP PREP to source node with path information.

- Element ID—TBD
- Length—Length of the IE
- Destination address—Address of destination MP
- Source Address—Address of originator MP
- Path information—Information about root node, all intermediate nodes, and destination address
- TTL—Time to live is the maximum numbers of hops allowed for this IE
- Metric—The cumulative metric from the originator to the MP transmitting the PREQ
- Life Time—The time for which the MPs receiving the PREQ consider the forwarding information to be valid.

- Modified Proactive Routing Mechanism

In proactive routing protocol, MPP will announce common neighbors of a source destination pairs as root in the networks. After receiving the neighbor's information, every node searches for the common neighbor between source S and destination D. If it has any common neighbor, then it announces that node as a root. If more than one common node is found between S and D, then the MPP tries to search for the less weighted node as the root. If both then common nodes have the same weight, then MPP announces the root on the basis of its priority.

If S changes in the networks, then new source node S always sends PREQ through a new path to the MPP. Subsequently, the MPP provide a new path according to the neighbors of S and D. The complete details of that path are stored in the second routing table for future use.

When any new node is added to the networks, then the new node just sends a hello message to the neighbors. The neighbors store information about that new node in the table and send the details to MPP. Next, the MPP updates its own table.

MPP maintains two tables. The first table stores information about the source and destination and about the neighbors of each node. The second table stores the path information between every node, and this table is maintained only after the first table. In the second table, the path information is stored until and unless there are any changes in networks. For example, when any nodes are deleted or added, then the MPP updates the second routing table according to changes in the networks.

Every MP maintains its table by providing information about the node id, neighbor's details, and path details (If those MPs play any role between the source and the destination, then the details of path is maintained by the MPP). The information element of RANN is defined as shown in Fig. 5.5.

- Element ID—TBD
- Length—Length of IE
- TTL—Time to live is the maximum numbers of hops allowed for this IE
- Metric—The cumulative metric from the originator to the MP transmitting the PREQ
- Life Time—The time for which the MPs receiving the request consider the forwarding information to be valid

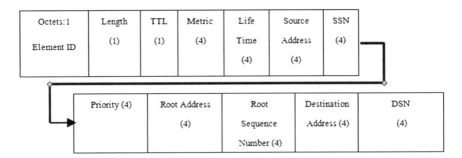

Fig. 5.5 Root announcement elements

- Source Address—Originator node whose need path for data transmission
- Root address—An address of a node that is announced as the root between the source and destination by the MPP
- Destination Address—The destination address of the source node.

We explain our proposed RANN and path selection mechanism by using the following example.

As shown in Fig. 5.6, let us consider networks that have a total of 8 MPs that are connected to one MPP. Node S is the source and node D is the destination.

MP S broadcasts PREQ to MPP (S-A-MPP) with information of destination id and neighbors' id (neighbors of S). Neighbors of 'S' are 'F', 'A', and 'C'. MPP stores all the information in the first table. Figure 5.7a shows the PREQ from S to MPP.

When MPP receives the PREQ from source with source id, destination id, and source neighbors id, it generates the neighbors query. Then, the MPP broadcasts the Neighbors query in the networks. As shown in Fig. 5.7b, the Neighbors query is broadcasted to all MPs.

When MPP sends the Neighbor query to all the seven nodes 'A', 'B', 'C', 'D', 'E', 'F', 'G', and then all the 7 MPs send a neighbor reply with neighbors information to MPP.

'A' replies 'S', 'C', and 'B'; 'B' replies 'A', 'C', and 'E'; 'C' replies 'A', 'B', 'S', 'F', 'E', and 'G'; 'E' replies 'B', 'C', and 'D'; 'F' replies 'C', 'S', and 'G'; and 'G' replies 'F', 'C', and 'D'.

When MPP receives all the neighbors information of all the MPs from each of the 7 nodes, it tries to find out the common node between S and D as follows:

Fig. 5.6 Wireless mesh networks topology

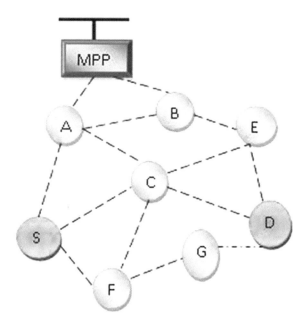

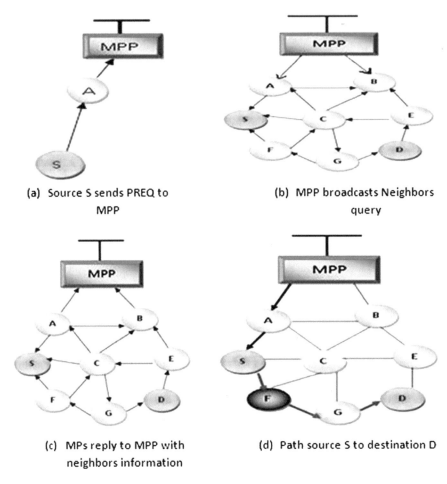

(a) Source S sends PREQ to
MPP

(b) MPP broadcasts Neighbors
query

(c) MPs reply to MPP with
neighbors information

(d) Path source S to destination D

Fig. 5.7 DHWMP path setup process

- 'S' neighbors—'A', 'F', and 'C'.
- 'D' neighbors—'G' and 'E'.

In this case, common neighbors are not found. Once again, the MPP attempts to find out the common neighbor amongst the source neighbors and destination neighbors.

MPP checks the neighbors of the source's neighbor

- (S-A)'s neighbors of A—S, C, and B.
- (S-F′)'s neighbors of F—S, C, and G.
- (S-C)'s neighbors of C—S, F, G, A, and B.

The PREP to MPP from all the MPs with neighbors' information is shown in Fig. 5.7c.

The MPP also checks for the destination neighbors

– (D-G)'s neighbors—F, C, and D.
– (D-E)'s neighbors—D, C, and B.

Next, the MPP compare all the neighbor nodes of source and destination neighbors to find out a common node.

The neighbors of A are compared with the neighbors of G—common neighbor is C.

The neighbors of A are compared with the neighbors of E—common neighbors are C and B.

Note: Here both C and B MPs are common neighbors of MP A and E, but MP B does not connect with S MP. Therefore, MPP chooses MP C as the common node.

The neighbors of F are compared with the neighbors of G—common neighbor is C.

The neighbors of F are compared with the neighbors of E—common neighbor is C.

The neighbors of C are compared with the neighbors of G—common neighbor is F.

The neighbors of C are compared with the neighbors of E—common neighbor is B.

Therefore, C and F are found to be the nearest neighbors between the source and the destination. MPP announces F as root by sending RANN to the source and destination because F is the most feasible node between S and D. F has less connectivity when compared with C. If C is declared as root, then traffic congestion may occur in future communication.

After sending RANN, MPP finds the final path and send PREP to the source. MPP stores the final path information such as the source id, destination id, and root id in the second routing table for future use.

When MPP received the shortest response from neighbors then MPP assign that shortest path for node to node approach for data transmission between S-F-G-D. After receiving the path, the source updates the information in its proactive routing table. When the on-demand routing protocol is used between S and D, S checks the proactive routing table, finds the path from the table, and sends a hello message to the final path. If the nodes F, G and D are found to be free, it informs to the source node and the data is transmitted. The final path is shown in Fig. 5.7d.

• Modified reactive routing mechanism

When a source MP, S needs a path to the destination MP, D, and then the pre paths of roots needs to be checked, which are already stored in the table. A pre path is created with the help of proactive path selection and RANN. If any source MP needs a route for its destination, then first, the source MP checks its proactive table, finds its path which was provided by MPP, and stores it in the source MP. Next, the source node sends the hello message to its nearest node. If a node is in sleeping mode, then after receiving the hello message, it gets activated and is ready for the response to the node according to message. If any node is busy with any other transmission, then that node receives information that one more action is waiting to be processed. Airtime link metric is always initialized to 0 in reactive routing. Whenever a MP forwards a hello message, the MP updates the metric field to reflect the cumulative metric of the path. After activating the path of the source node, the destination MP unicast PREP back to the source node.

When the source receives the PREP, it transmits data frames to the destination. If the destination receives further hello messages with a better metric, then the destination updates its path for the new path and sends the PREP to the source along the updated path. A route is considered to be active as long as there are data packets traveling periodically from the source to destination along the path described by the MPP [4]. It's consumed less energy compared with an existing PREQ message. This helps in saving power.

5.4   Performance Analysis of DHWMP Mechanism

We have use the simulation scenarios of NS-2, to analyze the performance results on the aggregate throughput are shown below. The results have been simulated over the 1000×1000 m^2 areas. We employed a constant bit rate (CBR) source for traffic scenario. The packets are 512 bytes' length and generated in 4 packets per second. The following parameters are used to simulate proposed DHWMP mechanism (Table 5.1).

These performance parameters have great impact to measure on overall performance of wireless network. In the different simulation scenarios has be studied according to four different performance metrics: channel capacity, packet delivery ratio, end-to-end delay, and routing overhead. In this scenario, performance of DHWMP protocol is evaluated in addition to OLSR, AODV and HWMP as a function of number of nodes. Figures 5.8, 5.9, 5.10, 5.11 and 5.12 depict the channel

Table 5.1 Parameters metrics for DHWMP [4]

Parameter	Value
Simulation area	1000 * 1000 m^2
Total no. of nodes	100
Node placement	Uniformly
Simulation time	200 s
Packet size	512 bytes
Bandwidth	2 Mbps
Traffic type	CBR
Packet rate	10 pkts/s

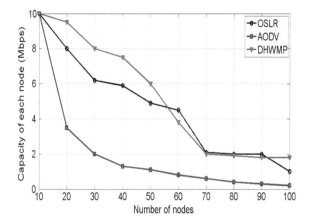

Fig. 5.8 Comparison of each node capacity into the WMN

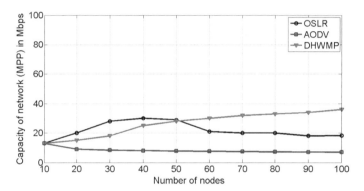

Fig. 5.9 Compression of the MPP capacity into the WMN

capacity, packet delivery ratio and the end to end packet delay, and routing overhead.

5.4.1 Channel Capacity

Figure 5.8 shows the performance comparison of the whole WMNs with respect to routing protocols (AODV, OLSR and DHWMP). The HWMP has given better performance compared to OLSR and AODV. In the beginning of the simulation, all three routing protocols start from the same point with fewer nodes, but when number of nodes increased, routing protocol's capacity performance went down due to network congestion and collision. But DHWMP still gives better performance compared to AODV and OLSR. When the number of node is 30, the difference

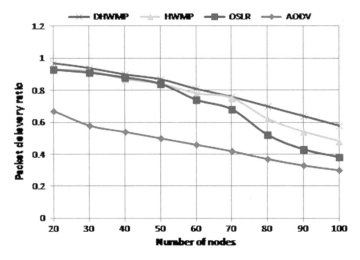

Fig. 5.10 Packet delivery ratio as a function of number of nodes

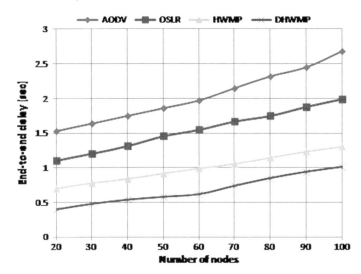

Fig. 5.11 End to end delay as a function of number of nodes

between DHWMP and AODV is the highest, i.e. almost 4.5 Mbps. Number of node 30 and 70 have respectively maximum (almost 2 Mbps) and minimum (0.01 Mbps) difference between OLSR and DHWMP.

The simulation result in Fig. 5.9 has shown the comparison of channel capacity performance for proposed DHWMP and other existing routing protocols (AODV and OLSR) for the whole networks. In the figure, we can see the differences between the routing protocols. When simulation has started all routing protocols' capacities were the same but as the numbers of nodes are increasing, AODV's performance

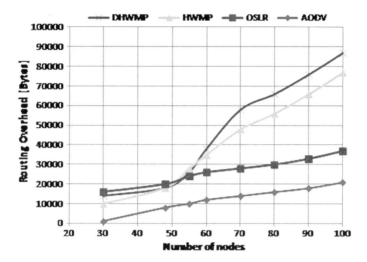

Fig. 5.12 Routing overhead as a function of number of nodes

went down but OLSR and DHWMP capacities increased. In the result, till number of 50 nodes, DHWMP performance is less than OLSR. After the number of 50 nodes, OLSR's performance went down compared to DHWMP. In a large scale network DHWMP gives better capacity performance compared to OLSR and AODV.

5.4.2 Packet Delivery Ratio

The packet delivery ratio is the ratio of received packets by destination and sent packets by source. It reflects the efficiency of the routing protocols.

Figure 5.10 show the packet delivery ratio of DHWMP, which is higher than HWMP and OSLR even AODV, has lowest performance. Although there is a better performance using single channel, the more the nodes transmitting packets through the network, the performance starts degrading. However, from Fig. 5.10, we can see that packet delivery performance of DHWMP with the increase in number of nodes is better than AODV, OLSR and HWMP. With 100 nodes, DHWMP has a packet delivery ratio of 59% whereas, HWMP OLSR and AODV have packet delivery ratio 44, 39 and 25% respectively. The simulation result confirms better performance of DHWMP over HWMP, OLSR and AODV.

5.4.3   End-to-End Delay

Figure 5.11 presents end to end delay with respect to DHWMP, HWMP, OLSR, and AODV. In this result, DHWMP has the lowest end to end packet delay compared to existing routing protocols. Figures 5.11 show the delay that occurs 0.4 s for 20 nodes. After that it slightly increases with respect to nodes, when the number of nodes becomes over 100. The ETE delay of DHWMP is 1 s, whereas AODV, OLSR and HWMP ETE delay are 2.7, 2, and 1.8 s respectively. Hence DHWMP has performed better than other routing protocols in terms of ETE delay also.

5.4.4   Routing Overhead

The number of routing packets requires by the routing protocol to construct and maintain routes. WMNs are scalable network. It measures how much network is scalable for this it necessary to measure the routing overhead of the network.

Figure 5.12 points out that the routing overhead of all DHWMP has less overhead compared to OLSR till 55 nodes. After 55 nodes HWMP and DHWMP's overhead to OLSR till 55 nodes. After 55 nodes HWMP and collision, but still HWMP's overhead is less than DHWMP. AODV, HWMP and OLSR have less overhead compare to HWMP. When numbers of nodes are a few then DHWMP gives little better performance compare to OLSR. DHWMP and HWMP have almost same overhead from node 48–56 nodes.

In results we can see that DHWMP and HWMP have a lot of overhead between 55 to 100 nodes. DHWMP has almost 10,000 bytes' difference to HWMP but compared to OLSR and AODV, huge bytes overhead occur. In result we can see, DHWMP has a lot of overhead in large scale networks.

5.5   Discussion

The HWMP provides a base and creates the easiest way to communicate with new emerging WMN technology. HWMP is a combination of reactive routing and proactive routing protocols. We have shown the performance of the DHWMP with AODV, OLSR and HWMP. We have measured the performance of DHWMP with respect to channel capacity, packet delivery ratio, end to delay, and routing overhead. DHWMP has given better performance for channel capacity, packet delivery ratio, and end to end delay except for routing overhead. Total cost of routing overhead is higher in our proposed DHWMP compared to others routing protocols. This is only drawback of DHWMP. It performs nicely in WMNs compare to other existing routing protocols.

5.6 Summary

HWMP routing protocol is the default routing protocol for wireless mesh networks. In HWMP, both reactive and proactive techniques are used, but there are so many drawbacks in the existing systems. Therefore, we have tried to solve some of the problems in this study. For example, the existing systems have root constraints and tree-based communication, which indicates that there is wastage of power and resources. To resolve this problem, we provide the decentralized root. To enable the communication between two nodes, it is not necessary that the same nodes should play the role of the root.

This proposed system shows different root techniques for different data transmissions; all these transmissions are carried out using the proactive routing protocol. In the existing HWMP, during reactive routing, if any node transmits the data, it's initiated with PREQ message. However, in this study, the node simply sends a hello message to initiate data communication. Therefore, we confirm that our proposal improves the drawbacks of the existing HWMP and presents a better HWMP [3].

References

1. M. Bahr, Update on the hybrid wireless mesh protocol of IEEE 802.11. IEEE International Conference on Mobile Ad-hoc and Sensor Systems, Pisa, Italy (2007)
2. M. Bahr, G. Strutt, W.S. Conner, Path selection metric framework. IEEE P802.11 Wireless LANs, document IEEE 802.11-07/0239r1 (2007)
3. M. Singh, S.-G. Lee, Decentralized hybrid wireless mesh protocol. ICCIT 2009, Seoul, ACM 978-1, pp. 824-829 (2009)
4. M. Singh, S.-G. Lee, H.J. Lee, Non-root-based hybrid wireless mesh protocol for wireless mesh networks. Int. J. Smart Home 7(2) (2013). ISSN No: 1975–4094
5. A.O. Lim, X. Wang, Y. Kado, B. Zhang, A hybrid centralized routing protocol for 802.11s WMNs, Special Issue on Springer, SI Advanced Wireless Mesh Networks, Mobile Networks and Applications (MONET), Netherlands (2008)

Chapter 6
Wireless Mesh Networks: Real-Time Test-Bed

6.1 Test-Bed Setup

In this section we defined the specification of our test-bed setup requirements. We have used Ubuntu Linux with kernel 2.6 based PCs and Multiband Atheros Driver for Wi-Fi (MADWIFI) diver for virtual access point (VAP) creation. The VAP has been created for the operating system as different wireless interface. Moreover, every VAP can be made for different mode. The MADWIFI driver creates VAP interfaces for Linux Bridge. That bridge works same as Hardware Bridge. The port of the bridge is communicated with each AP and wireless distributed system (wds), and those bridges forwards the frames between interfaces according to the destination MAC address. The implementation specification is described below [1].

6.2 Requirements Specifications

6.2.1 Hardware for Mesh Portal & Mesh Access Points

- Wireless NIC: Atheros communications Inc. AR5413 802.11abg.
- Wired NIC: Realtek Semiconductor Co., Ltd. RTL8111/8168B PCI Express Gigabit Ethernet.
- Wireless NIC: Atheros AR9285 Wireless Networks Adapter (Mobile Station).

6.2.2 Software Mesh Portal, Mesh Access Points & Mobile Station

- OS: Ubuntu version 10.04.2, kernel 2.6.32-31-generic.
- Driver: Madwifi, ath_pci.
- Measuring Tool: Jperf (Jperf is a graphical user interface using Iperf, which allows us to calculate the throughput results) [3].
- NIC driver: ath9k.

© The Author(s), under exclusive license to Springer Nature Singapore Pte Ltd., part of Springer Nature 2019
M. Singh, *Node-to-Node Approaching in Wireless Mesh Connectivity*, SpringerBriefs in Applied Sciences and Technology, https://doi.org/10.1007/978-981-13-0674-7_6

6.3   Operating System

In our experiment, we have used Ubuntu operating system. The Ubuntu is an operating system which is made by worldwide expert developers. The Ubuntu is open source, anyone can contribute to Ubuntu project by writing new software, packaging additional software, or fixing bugs in existing software. It's freely available and anyone can download from internet.

For the experiment, we have used Ubuntu version 10.04.2 because it was the latest version when this project started. The Ubuntu has its own wireless driver. But that wireless driver does not support WMNs. So, we need to install Multiband Atheros driver for Wifi (MADWIFI) [2].

6.4   Internetworking

Being the mesh completely at level-2, all higher-level protocols such as IP and ARP are supported in the conventional mode.

Typically, the portal interface is connected to an Ethernet LAN, and thus the node can be configured in two ways.

Router: the node is a networks termination point; all incoming and outgoing frames are handled at the IP level. This will break the LAN from outside the mesh [3] (Fig. 6.1).

6.5   MPs Communication (Broadcast)

In the architecture, the communication take place on the wds links, while making use of four addresses, and effect point to point links. In creating a wds link, the MAC

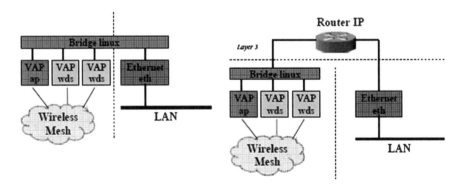

Fig. 6.1 Internetworking structure [3]

address of the node must be specified at the other end of the link. Yet, broadcast address cannot be used in creating a VAP in wds mode [4].

Hence the level-2 broadcasts on this architecture could only be achieved by transmitting a frame on all wds interfaces in separated accesses to the medium.

In theory, we could send a frame on a wds link that addresses broadcast destination. In reception, the Ethernet frame is extrapolated from the 802.11 frame, and then it enters into the bridge, which handles the broadcast Ethernet packet in traditional way is sent to all other active interface. At this point each wds interface transmits the frame over the wds link to which it is associated. Since this process is repeated at every MP, a loop may be easily created [3].

The problem of creating broadcast loops is shown in the Fig. 6.2 suppose S sends a broadcast packet that is received by MAP3. The bridge forwards the frame on both Ethernet interfaces wds1 and wds2. These frames will be sent at two different times by the driver. The bridge of MAP2, in turn, forwards the frame on all ports except the one from which the bridge has received it. It also forwards to MAP1, which had already received the frame from MAP3. MAP1 in turn retransmits the frame to MAP3, creating a loop. We preferred to find a solution that limited the use of broadcast.

We distinguished between broadcast from the Root to leafs (MPs) of the tree and from the Leafs (MPs) to the Root. The broadcast is typically used to reach all nodes. So we added a control to force MP to accept broadcast frames only from the parent in the tree.

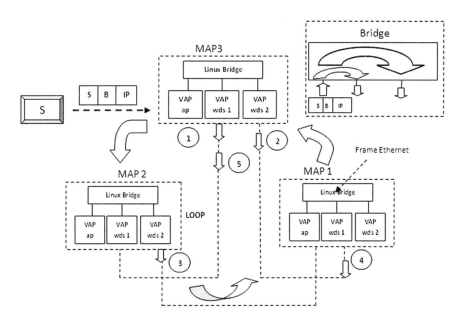

Fig. 6.2 Loop problem during broadcast

6.6   Routing Process

Once the mesh software is started, mesh portal will start to form the tree for the mesh networks. The tree is formed on the basis of the metric value or the radio quality of each link. Since the mobile station is not the part of the mesh networks, it was identified as a proxy station connected to the mesh access point. The mesh portal gains information about the path metric by broadcasting the path request frames to every mesh access points within the mesh networks. Every MAPs will then reply to the request and paths between the stations and the portal which are formed [5].

   The reactive routing is triggered by the mesh software after the proactive tree is formed, so that the path formation can be observed. The formation of reactive paths for this test is very similar to the formation of proactive paths. Path request frames will be unicast to every MAP that it can detect and the MAPs will reply to the request.

6.6.1   Path Formation [3]

- Path request (PREQ)

The main function of this frame is to let all the stations that it past to learn about the next hop MAC address given the destination MAC address. The operation of PREQ is shown in Fig. 6.3a by assuming that a Mesh STA A initiated the path formation. The PREQ is assumed to past through two intermediate stations before reaching the destination.

- Path reply (PREP)

The operation is similar to the PREQ where the destination station will reply to the

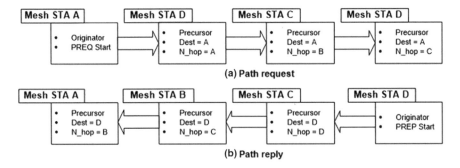

Fig. 6.3 Path communications in networks [30]

originator station. The operation is shown in Fig. 6.3b. When the PREP frame reaches the originator station, a bidirectional path is formed.

- Path selection

In case of multiple paths, the best path must be chosen. This is done by comparing the metric value of the path. A lower value means that the path is better therefore the path with the lowest metric will be chosen.

6.6.2 Metric Formula

Research has been done by Garroppo et al. [6] stating that the default airtime link metric causes the metric value to change very often and they had proposed their own metric formula. The test is done in our lab by using the exact metric formula.

The formula is shown in Eq. 6.1, where S is the signal strength and ε_r is the error rate.

$$metric_{Link} = \left[\frac{2}{1-\varepsilon_r}\right]\left[\frac{125}{S} + 1.6\right] \tag{6.1}$$

In the case of multiple paths, the best path must be chosen. This is done by comparing the metric values of the path. In Fig. 6.4 we can easily see that all path metric values from Mobile 1 to Mobile 2.

Different metric values are shown in below.

- Mobile 1–Mesh 2–Mesh 4–Mobile 2 = 30
- Mobile 1–Mesh 2–Mesh 1–Mesh 4–Mobile 2 = 20
- Mobile 1–Mesh 2–Mesh 3–Mesh 4–Mobile 2 = 30

The lowest path metric value Between Mobile 1 to Mobile 2 is Mobile 1–Mesh 2–Mesh 1–Mesh 4–Mobile 2 = 20. The final best path selection is shown in Fig. 6.5.

6.7 Networks Setup

In this part, we discussed an example of our experiment. In our experiment, we used 4 PCs with mesh features as MP and 2 laptops (no mesh) as station. Each system has been assigned a unique name, and IP address. Every system has its own MAC address. Test bed layout is shown in Fig. 6.6.

In this section, we mentioned an example based on real implementation test. During implementation, we used 1 mesh portal point (MPP), 3 mesh point (MAP1, MAP2 & MAP3), and 2 clients (sta1 and sta2). The tree was created after the initialization (MPP-Mesh 2—mesh 4 and MPP-mesh 3). The application required to get the MAC address of mobile station 1 and station 2, whose IP address is known.

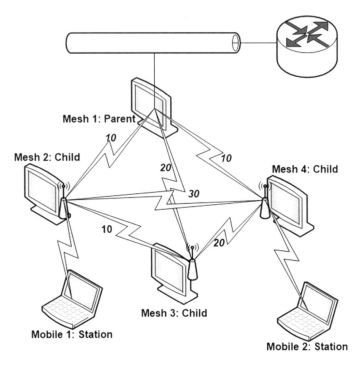

Fig. 6.4 Path metric between nodes [30]

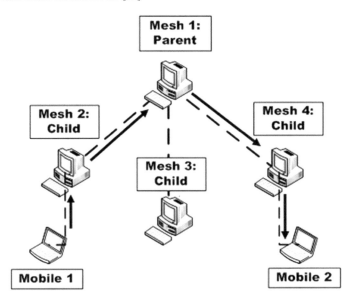

Fig. 6.5 Best path selections [30]

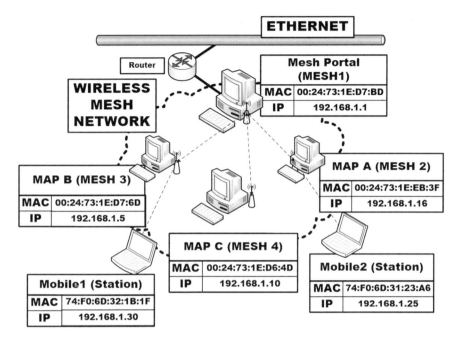

Fig. 6.6 Test bed wireless mesh networks infrastructure [30]

The mobile station 1 has broadcast path request (PREQ) message, which received by MAP4. This node unicast PREQ to final destination the mesh portal point (MPP). The message frame was thus sent to MAP2 which forwards it to MPP. MPP has information about all stations of the networks and replied to sta2's MAC address. The ARP message replied to the address of mobile station 1 that reached MAP2, then MAP4 and then the final destination i.e. mobile station 2. We can see the nodes IP address and MAC address in Table 6.1.

Once, the path has been created between mobile station 2 and 1 with IP-MAC address of mobile station 1. It starts transmitting data frames to MAP2 with final destination mobile station 2. When mesh 4 receive frames without any route to

Table 6.1 Computer information (ID, IP address, MAC address)

	Station	Mac address	IP address
Mesh 1	Portal/Root station	00:24:73:1E:D7:BD	192.168.1.1
Mesh 2	Mesh station	00:24:73:1E:EB:3F	192.168.1.16
Mesh 3	Mesh station	00:24:73:1E:D6:6D	192.168.1.10
Mesh 4	Mesh station	00:24:73:1E:D7:4D	192.168.1.20
Mobile 1	Mobile station	74:F0:6D:31:23:A6	192.168.1.25
Mobile 2	Mobile station	74:F0:6D:32:1B:1F	192.168.1.30

mobile station 2, it could forward the PREQ to MPP. The path created with the help of ROOT and data transmits to MAP3 and then mobile station 2.

The Mesh 4 sends PREQ message through broadcast manner to setup a direct path to mobile station 2, then data is transmitted to the mesh portal. When PREQ message received by Mesh 3, it acts as proxy of mobile station 1 and sends the path reply (PREP) to Mesh 4. After receiving PREP Mesh 4 start data transmission via built path (in our case MAP4-MAP2-MAP3).

6.7.1 Real Position of Stations

See Fig. 6.7.

6.7.2 Real-Time Test-Bed Results

Tree

===================================

Time: 40.100745

Parent: ->00:24:73:1E:D7:BD
Child: ->00:24:73:1E:EB:3F
Child: ->00:24:73:1E:D6:6D
Child: ->00:24:73:1E:D7:4D

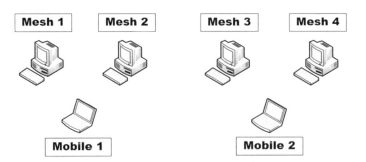

Fig. 6.7 Real position of the stations [30]

```
===================================
```

ROUTE TABLE (MESH 1)

```
===================================
```

Time: 16.82198

DESTINATION: 00:24:73:1E:D6:6D
NEXT HOP: 00:24:73:1E:D6:6D
SEQUENCE NUMBER: 0
HOP COUNT: 1
PATH METRIC: 302
Flag TYPE: 1
Flag SELF ROUTE: 0
Flag ROUTE_VALID: 1
Flag NEIGH: 1

```
===================================
```

DESTINATION: 00:24:73:1E:D7:4D
NEXT HOP: 00:24:73:1E:D7:4D
SEQUENCE NUMBER: 0
HOP COUNT: 1
PATH METRIC: 302
Flag TYPE: 1
Flag SELF ROUTE: 0
Flag ROUTE_VALID: 1
Flag NEIGH: 1

```
===================================
```

----THIS IS SELF ROUTE
DESTINATION: 00:24:73:1E:D7:BD
NEXT HOP: 00:24:73:1E:D7:BD
SEQUENCE NUMBER: 2
HOP COUNT: 0
PATH METRIC: 0
Flag TYPE: 1
Flag SELF ROUTE: 1
Flag ROUTE_VALID: 1
Flag NEIGH: 0

```
===================================
```

DESTINATION: 00:24:73:1E:EB:3F
NEXT HOP: 00:24:73:1E:EB:3F
SEQUENCE NUMBER: 0
HOP COUNT: 1
PATH METRIC: 0
Flag TYPE: 1

Flag SELF ROUTE: 0
Flag ROUTE_VALID: 1
Flag NEIGH: 1

Table below shows how the metric for each path changes with time; for the Mesh 1 station.

```
=================================
```
Neighbor list (Mesh 1)
```
=====================================
```
Time: 20.85116
MAC ADDRESS: 00:24:73:1E:D6:6D
STATE: 1
INTERFACE: wdsX1
TX BIT RATE: 58
TX PACKET ERROR RATE: 992
RX BIT RATE: 0
RX PACKET ERROR RATE: 1
AVERAGE RSSI: 0
LINK METRIC: 100
LIFETIME: 6737 milliseconds

```
=================================
```

MAC ADDRESS: 00:24:73:1E:D7:4D
STATE: 1
INTERFACE: wdsX2
TX BIT RATE: 26
TX PACKET ERROR RATE: 905
RX BIT RATE: 0
RX PACKET ERROR RATE: 1
AVERAGE RSSI: 0
LINK METRIC: 100
LIFETIME: 11830 milliseconds

```
=================================
```

MAC ADDRESS: 00:24:73:1E:EB:3F
STATE: 1
INTERFACE: wdsX3
TX BIT RATE: 33
TX PACKET ERROR RATE: 977
RX BIT RATE: 0
RX PACKET ERROR RATE: 1
AVERAGE RSSI: 0
LINK METRIC: 100
LIFETIME: 14370 milliseconds

The metric is very unstable. Therefore, the metric calculation for the system is studied to figure out the formula used.

IF Metric <0 or Metric >100;

THEN Metric = 100;

Where error is the error rate and signal is the signal strength for the link. Changes are needed to make the metric more stable.

When the mobile station connected to the mesh networks, the root station is aware of the mobile station and the access point that the mobile station is associated.

6.8 Performance Analysis

6.8.1 TCP Throughput Analysis

The experiment had been done according to the experimental scenario described below. First, settings and parameters for TCP test were defined. Then an experiment has been initialized with a simple TCP test. We have done 5 trials for TCP protocol transmission. The information of the test is shown in Table 6.2.

We have done 5 trails of TCP transmission with different window size. We have set 2 MB for the buffer length in experiment. Length of buffer is set for the bandwidth read and write. By default, it uses 8 KB for TCP. We can also set the window size for TCP experiment. The window size is a socket buffer size to the specified value in TCP transmission. We have used different window size for every experiment. For window size we used 5 different values such as 16, 32, 128, 128, and 256 KB. We printed the reported TCP maximum segment size (MSS) and the observed read sizes which often correlate with the MSS. The MSS is usually the maximum transmission unit (MTU) 40 bytes for the TCP/IP header.

The interface type corresponding to the MTU is also printed (ethernet, etc.). This option is not implemented on many OS, but the read sizes may still indicate the MSS. We have set the MSS 1 KB in our experiment. We have run our transmission for 100 s of each trail and set 10 s interval time between periodic bandwidth.

Table 6.2 Parameters for TCP experiment

Transport protocol	TCP
Number of trials	6
Buffer length	2 MB
Max segment size	1 KB
Total transmission time	100 s
Interval	10 s
Transport Protocol	TCP
Number of trails	6

In Table 6.3 we can see average throughput results of the all 5 different window size. We can see the variation of throughput between different window sizes. In Fig. 6.8, we can see, the window size 128 KB throughput results is better than other window size throughput. TCP transfer much data when bandwidth is of 128 KB. In Fig. 6.9 we can see that data transfer rate of 128 KB is more as compared to others.

6.8.2 UDP Throughput Analysis (Throughput Analysis for Different Data Packets)

After TCP experiment, we have set the parameter for UDP test. UDP test parameter settings are shown in Table 6.4. The UDP bandwidth has been set to 200 KB/s at the client side thus the client will constantly send out data frames at the stated speed. The bandwidth of the path is determined by the datagram ratio where higher datagram ratio means a lower path bandwidth with a maximum bandwidth of 200 KB/s if no data is lost. For an experiment we have used 6 datagram sizes (500, 1000, 1500, 2000, and 2500 bytes). We have defined the common buffer size for experiment which is 41 KB. The buffer size is just buffer which packets are received in, and limits the maximum receivable data packets size. We have fixed 10 common transmission times

Table 6.3 Throughput results of TCP transmission

Window size (KB)	Bandwidth (KB/s)	Data transfer (KB)
16	2319	28,352
32	3224	39,368
64	3819	46,616
128	5677	69,312
256	3620	44,272

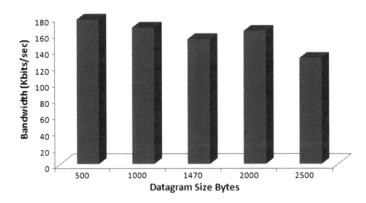

Fig. 6.8 TCP throughput result of bandwidth on different window size

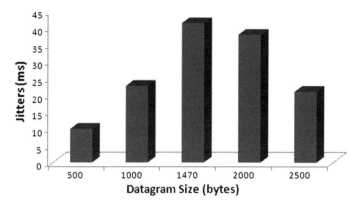

Fig. 6.9 Data transfer on different window size

for every trails. We tested every packet size to 3 times trails. Average throughput result of different packet size after experiment is shown in Table 6.5.

The result shows that jitters and datagram loss ratio increases with increasing datagram size. The jitters or packet delay variation and increased sharply when the datagram size exceeds the maximum size allowable (1470 bytes). Since the size is larger than 1470, the data packets have been fragmented and was sent separately. This causes the high latency as shown by the jitters. Surprisingly, the performance of the networks depends on the frame size but not the number of frames (Figs. 6.10 and 6.11).

Table 6.4 Parameters for UDP transmission

Transport protocol	UDP
UDP bandwidth	200 kbps
Buffer size	41 KB
Number of data transmission for each trail	10
Number of trails	3

Table 6.5 Throughput results of UDP transmission

Datagram size (bytes)	Bandwidth (KB/sec)	Jitters (ms)	Data loss rate (Loss/Total) (%)	Number of frames
500	187	15.659	273/5001 (5.5%)	5001
1000	172	25.525	327/2501 (13%)	2501
1500	185	9.365	122/1668 (7.3%)	1668
2000	163	37.998	228/1251 (18%)	1251
2500	183	21.084	83/1001 (8.3%)	1001

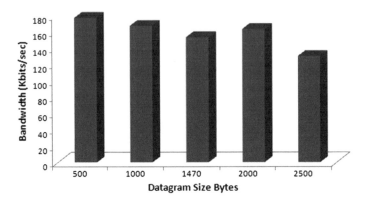

Fig. 6.10 Average throughput of different data grams of UDP

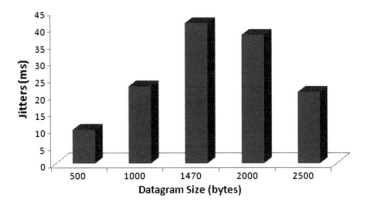

Fig. 6.11 Average jitter of each data grams of UDP

6.9   Summary

This chapter has presented an experimental performance analysis of different window size for TCP and UDP transmission. Most of the researchers did the wireless mesh networks performance analysis through the simulation. The simulation analysis is easier to be used but the result is often less reliable compared to real implementation. Therefore, a real implementation test is assumed to be able to provide a more reliable result.

Our test-bed is constructed on the basis of the IEEE802.11s draft 4.0. The networks so formed consist of four PCs and two laptops. In our experiment, the best path selection is done by using the modified air time metric which is claimed to be able to reduce the PREQ and RREQ frames flooding. In our experiment, the non-mesh stations information is stored through the proxy mechanisms in HWMP tree table whenever HWMP tree is available in the networks.

References

1. M.S. Siddiqui, C.S. Hong, *HRP: A Hybrid Routing Protocol for Wireless Mesh Network*, JCCI 2007 (Phoenix Park, Korea 2007)
2. Madwifi: multiband atheros driver for wifi [Online] Available: http://madwifi-project.org
3. M. Singh, S.G. Lee, W.K. Tan, J.H. Lam, Impact of Wireless Mesh Networks in Real-time Test-bed Setup. Adv. Inf. Sci. Serv. Sci. (AISS) (Scopus Journal) **3**(9), 25–33 (2011). ISSN No: 2233-9345
4. M. Segata, N. Facchi, L. Maccari, R.L. Cigno, RoRoute: Tools to experiment with routing protocols in WMNs, in *2018 14th Annual Conference on Wireless On-demand Network Systems and Services (WONS)*, Isola 2000, France (2018), pp. 91–94. https://doi.org/10.23919/wons.2018.8311668
5. N.M.A. Latiff, I. Ibrahim, S.K.S. Yusof, N.N.N.A. Malik, R. Arsat, A.S. Abdullah, Load distributed routing protocol for wireless mesh networks, in *2016 IEEE 3rd International Symposium on Telecommunication Technologies (ISTT)*, Kuala Lumpur (2016), pp. 41–46. https://doi.org/10.1109/istt.2016.7918082
6. R.G. Garroppo, S.Giordano, and L. Tavanti, Implementation frameworks for IEEE 802.11s systems. Proc. Comp. Commun. (Elsevier Journal) **33**, 336–349 (2010)

Future of Wireless Mesh Networks

The Wireless Mesh Networks are a revolutionary technology in 802.11 WLANs networks. It has become increasingly famous in last few years. WMNs have promising approach which enables the requirements of wireless networks and users in a flexible way. WMNs have reliable and self-configuring architecture. It is rapid and wide area deployment with self-organizing features. IEEE 802.11 groups are publishing the specification draft IEEE 802.11 groups are publishing the specification draft for WMNs but still it's not final draft. Due to this WMNs is open for research for all area such as architecture, routing, QoS etc. In this Book, we have presented description, including the implementation details with real-time experimental performance evaluation of node to node approaching data in WMNs. The experimental results show the throughput performance for different bandwidth and various data packets for data transmission. Routing techniques are also very emerging research field in WMNs even though HWMP is default routing protocol described for WMNs. The HWMP still has many drawbacks such as centralized route and energy constraints bottleneck problem.

In this book, we have discussed and shows the implementation of node to node appraocing data through decentralized hybrid wireless mesh protocol (DHWMP) and node to node cluster based data approaching. In these routing protocol we have tried to solve the above mentioned drawbacks. We have simulated and realtime implemented routing techniques through the NS-2 network simulator. We have shown the compative results of node to node to approaching mechanism in WMNs.

© The Author(s), under exclusive license to Springer Nature Singapore Pte Ltd.,
part of Springer Nature 2019
M. Singh, *Node-to-Node Approaching in Wireless Mesh Connectivity*, SpringerBriefs
in Applied Sciences and Technology, https://doi.org/10.1007/978-981-13-0674-7

Printed in the United States
By Bookmasters